Charts & Graphs

SURVEYING MATHEMATICS MADE SIMPLE

An original Book by

Jim Crume P.L.S., M.S., CFedS

Co-Authors
Cindy Crume
Bridget Crume
Troy Ray R.L.S.
Mark Sandwick P.L.S.

KINDLE - PRINTED EDITIONS

*****

PUBLISHED BY:

Jim Crume P.L.S., M.S., CFedS

Charts & Graphs

Copyright 2015 © by Jim Crume P.L.S., M.S., CFedS

All Rights Reserved

First publication: September, 2015

Cover photo: Near US 60 through Oak Flat Arizona

TERMS AND CONDITIONS

The content of the pages of this book is for your general information and use only. It is subject to change without notice.

Neither we nor any third parties provide any warranty or guarantee as to the accuracy, timeliness, performance, completeness or suitability of the information and materials found or offered in this book for any particular purpose. You acknowledge that such information and materials may contain inaccuracies or errors and we expressly exclude liability for any such inaccuracies or errors to the fullest extent permitted by law.

Your use of any information or materials in this book is entirely at your own risk, for which we shall not be liable. It shall be your own responsibility to ensure that any products, services or information available in this book meet your specific requirements.

This book is covered by the Kindle Direct Publishing and/or CreateSpace Terms and Conditions.

This book may not be further reproduced or circulated in any form, including email. Any reproduction or editing by any means mechanical or electronic without the explicit written permission of Jim Crume is expressly prohibited.

TABLE OF CONTENTS

INTRODUCTION 4
TYPICAL TOWNSHIP 5
TYPICAL REGULAR SECTION 6
STANDARD PARALLELS & GUIDE MERIDIANS.....7
PRINCIPAL MERIDIANS AND BASE LINES 8
SECTION - CORNER IDENTIFICATION....................9
NW QUARTER - CORNER IDENTIFICATION........10
NE QUARTER - CORNER IDENTIFICATION..........11
SE QUARTER - CORNER IDENTIFICATION 12
SW QUARTER - CORNER IDENTIFICATION.........13
THE LEGAL DESCRIPTION SYSTEM.......................14
PARTS OF A TYPICAL LEGAL DESCRIPTION........15
BOUNDS 16
METES AND BOUNDS 17
PLSS 18
CALL FOR ANOTHER DOCUMENT 19
"LY" 20
STRIP..21
LOT AND BLOCK 22
TANGENT DEFLECTION 23
DEGREE OF CURVATURE..24
ABOUT THE AUTHOR...25

Charts & Graphs

INTRODUCTION

Straight forward Step-by-Step instructions.

This book is just one part in a series of digital and paperback books on Surveying Mathematics Made Simple. The subject matter in this book will utilize the methods and formulas that are covered in the books that precede it. If you have not read the preceding books, you are encouraged to review a copy before proceeding forward with this book.

For a list of books in this series, please visit:

http://www.cc4w.net/ebooks.html

Prerequisites for this book:

A basic knowledge of Public Land Survey System (PLSS), writing Legal Descriptions, Centerline Tangent lines and Degree of Curvature are required for the Charts and Graphs shown in this book.

The following are several useful charts and graphs as a quick reference that show the layout of a Typical Township, the aliquot parts to a Typical Section, Standard Parallels, Guide Meridians, Principal Meridians & Base Lines, Corner Identification, stampings on monuments, the Legal Description System with examples, Tangent Deflection and when to show the bearings to a tenth of a second and the effects of the Degree of Curvature in regards to the chord distance vs arc distance.

Charts & Graphs

TYPICAL TOWNSHIP

Historical Tidbits

The Public Land Survey System (PLSS) was officially created by the Land Ordinance Act on May 20, 1785.
Sec. 1-1 & 1-20(n) (2009)

Versions of the "Manual of Surveying Instructions" 1855, 1871, 1881, 1890, 1894, 1902, 1930, 1947, 1973 and 2009.
Sec. 1-11 (2009)

School Section

1 - Order of Subdivisional Lines

Copyright (c) 2015

SURVEYING MATHEMATICS MADE SIMPLE
Apps - Books - eBooks
www.cc4w.net

Typical Township

5

TYPICAL REGULAR SECTION

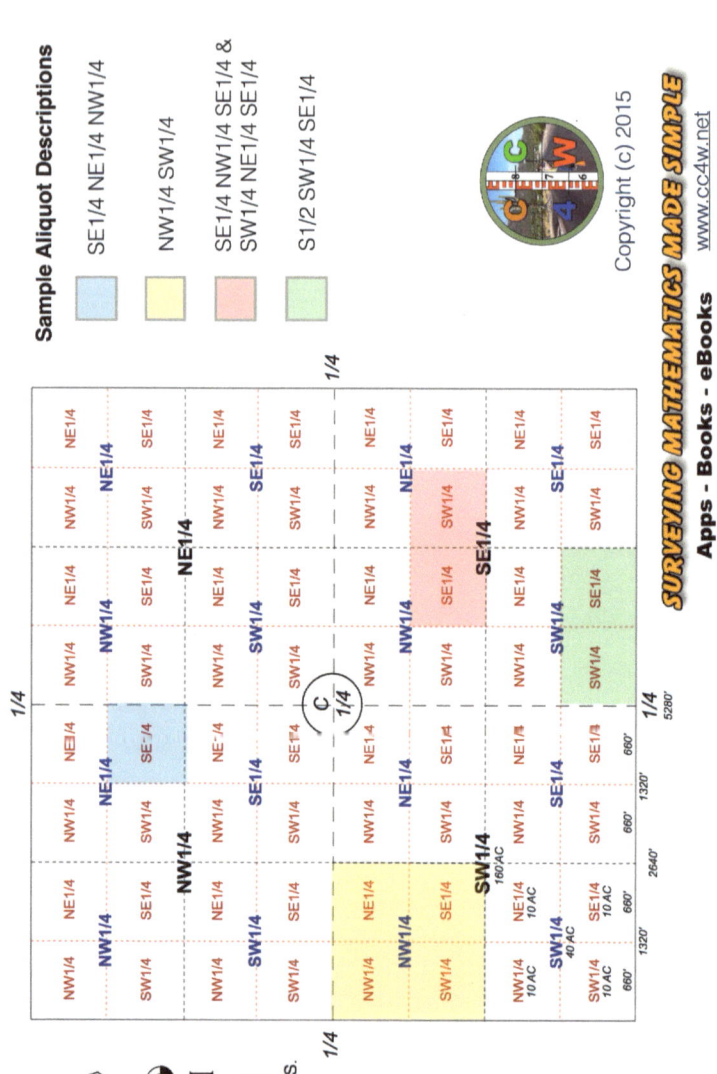

STANDARD PARALLELS & GUIDE MERIDIANS

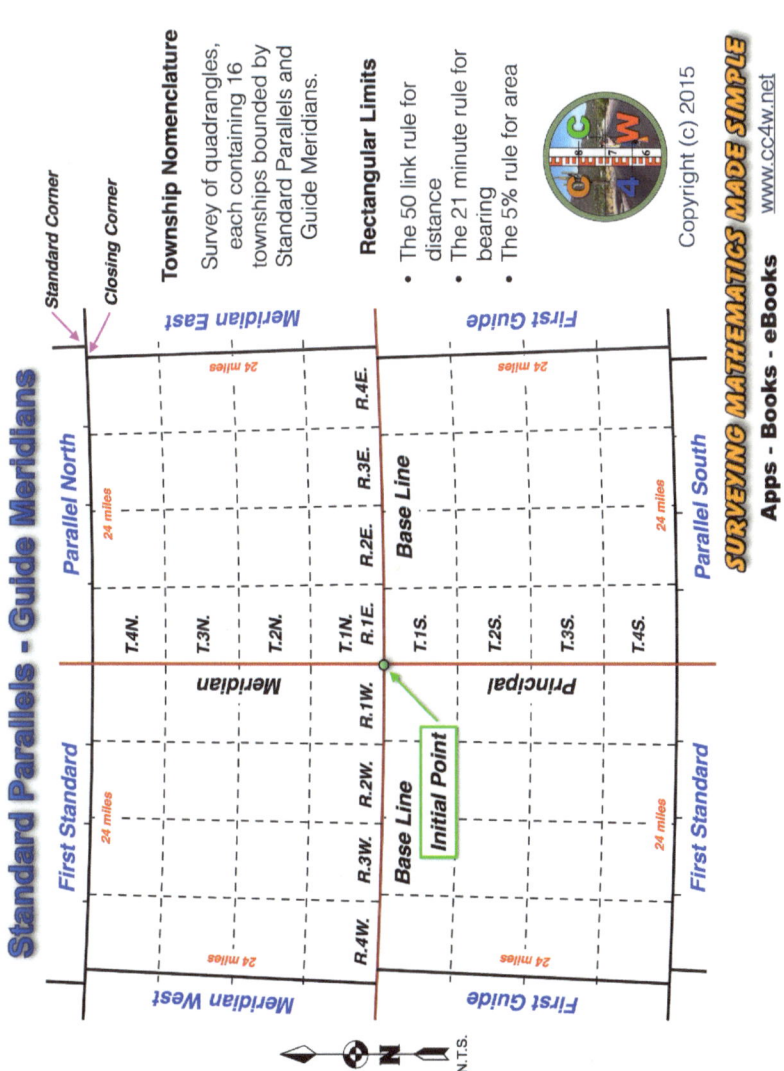

Charts & Graphs

PRINCIPAL MERIDIANS AND BASE LINES

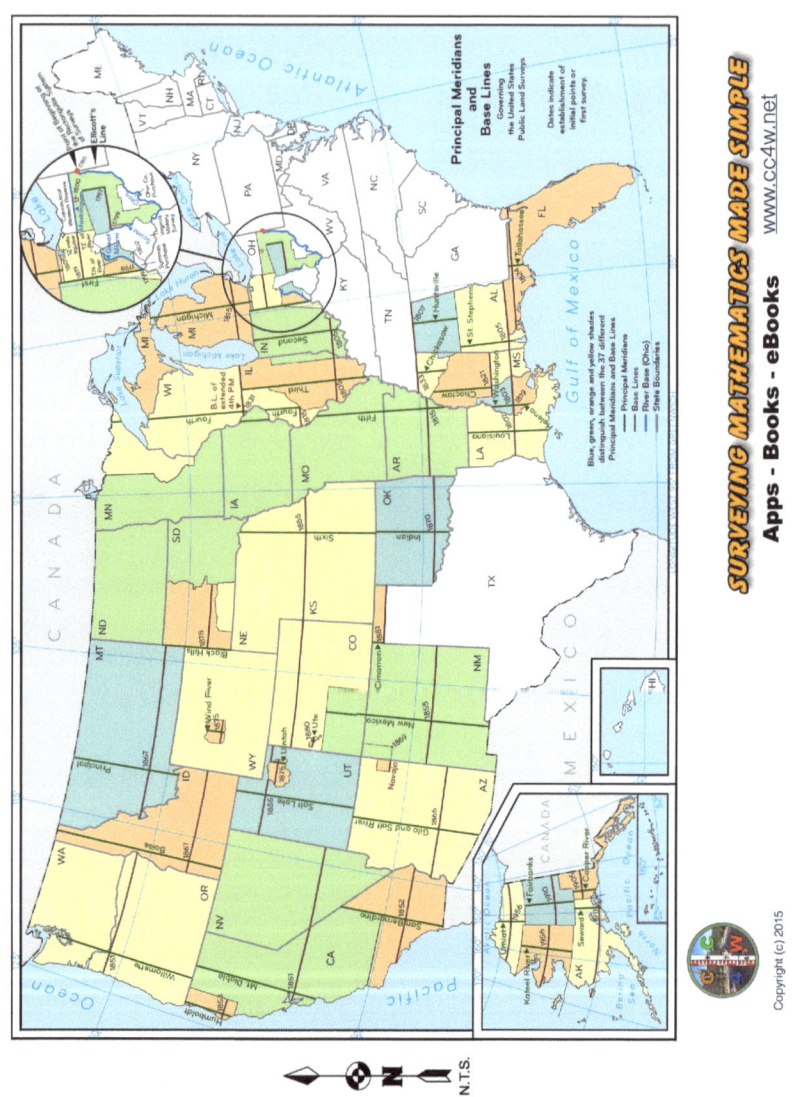

Charts & Graphs

SECTION - CORNER IDENTIFICATION

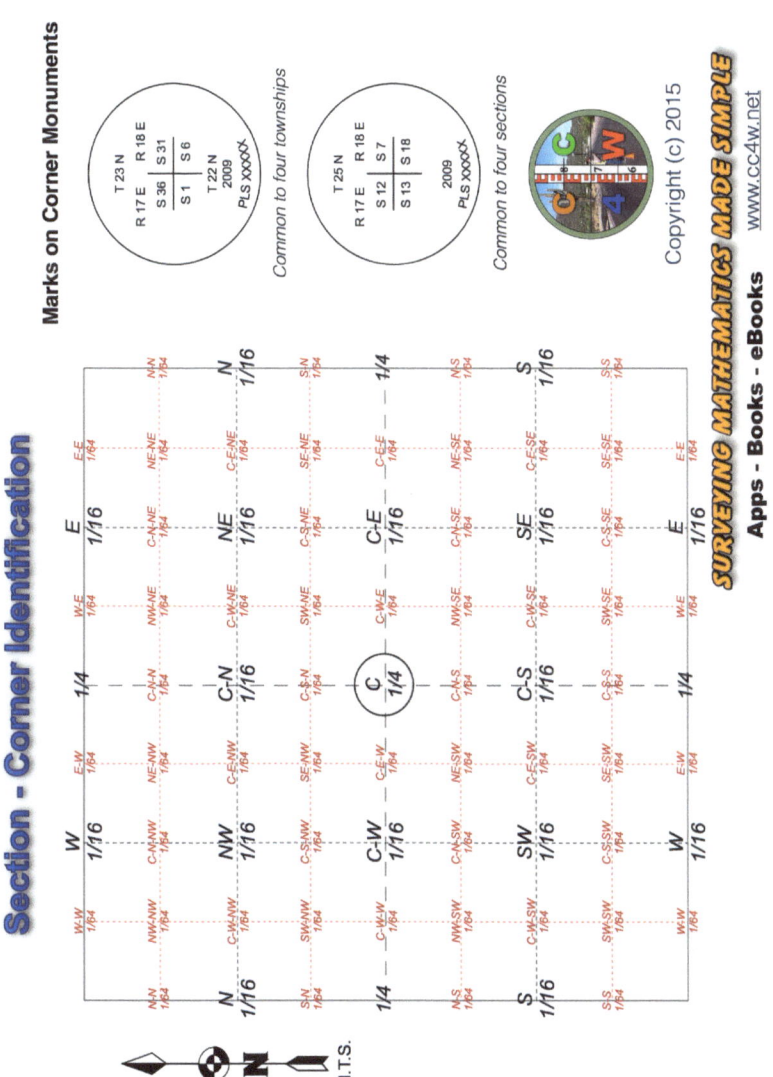

Charts & Graphs

NW QUARTER - CORNER IDENTIFICATION

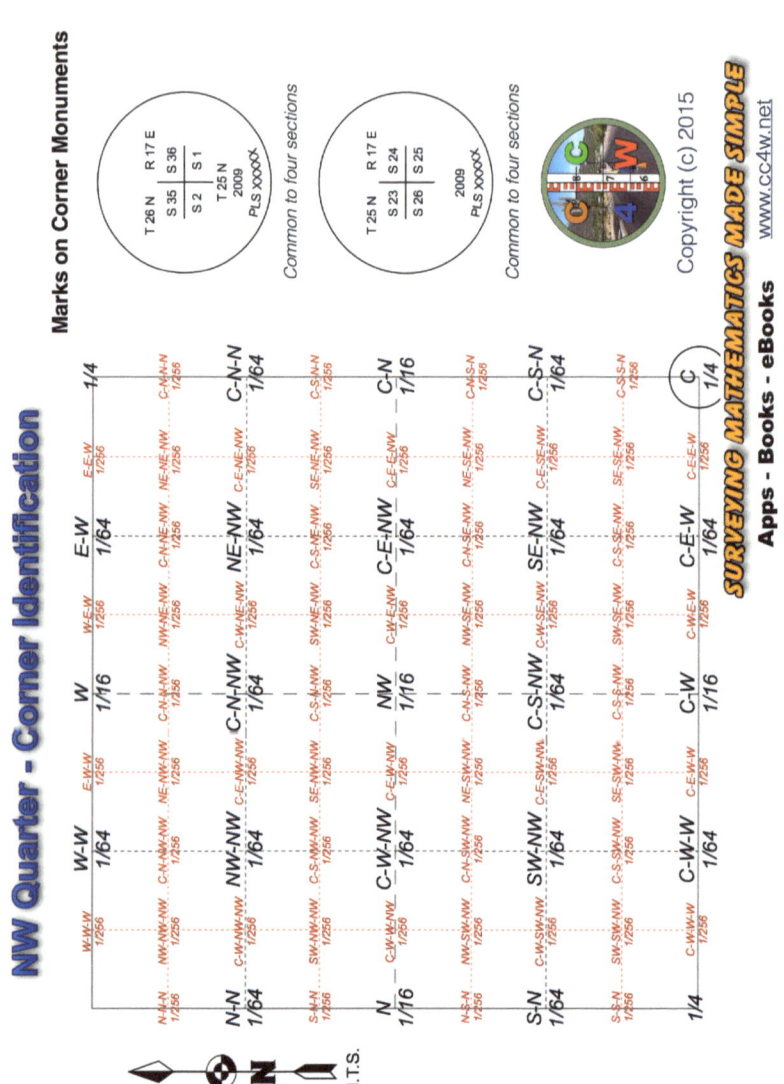

Charts & Graphs

NE QUARTER - CORNER IDENTIFICATION

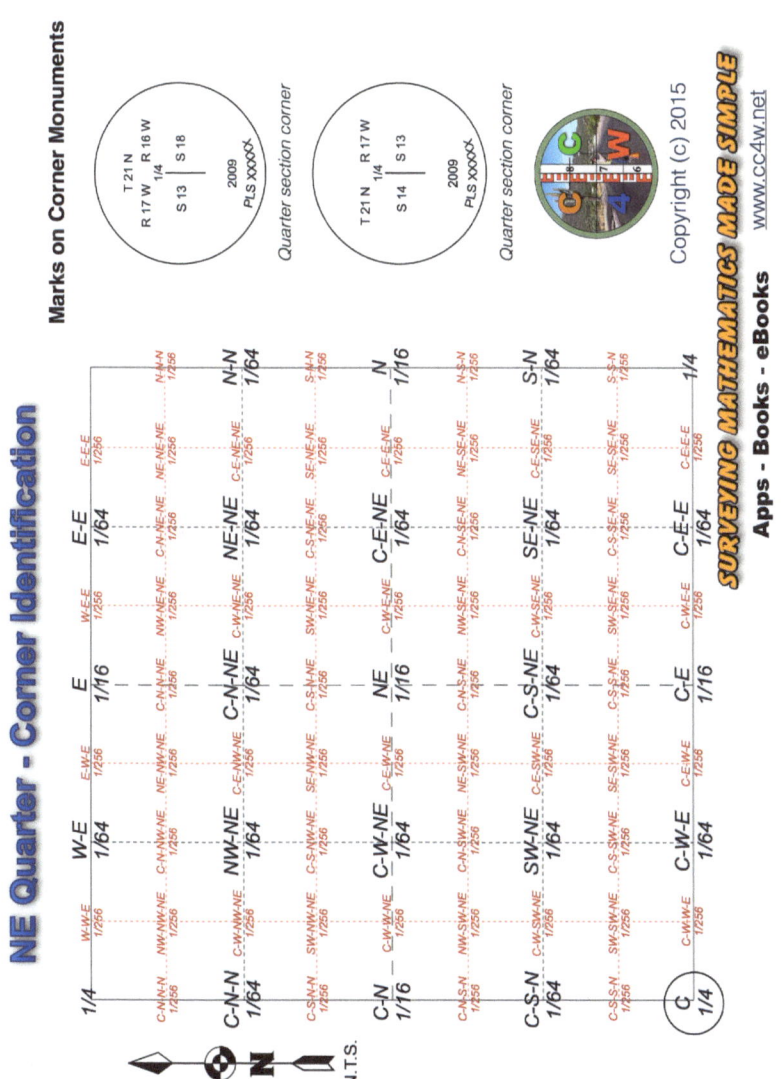

SE QUARTER - CORNER IDENTIFICATION

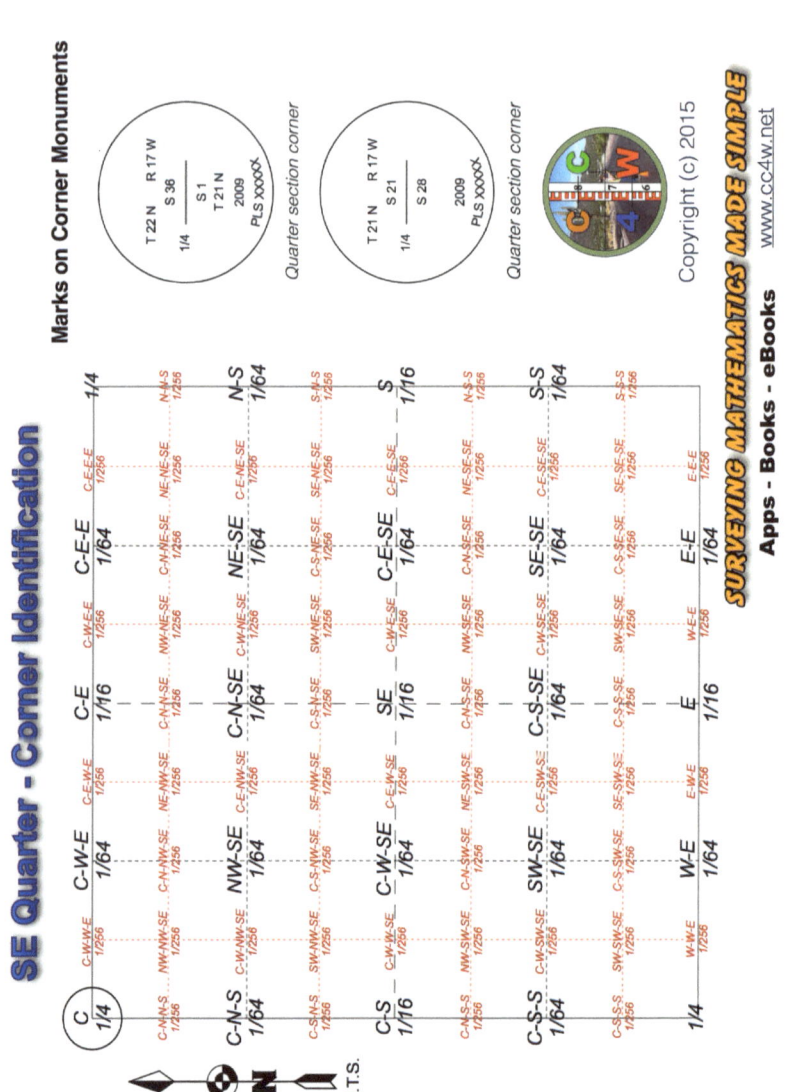

SW QUARTER - CORNER IDENTIFICATION

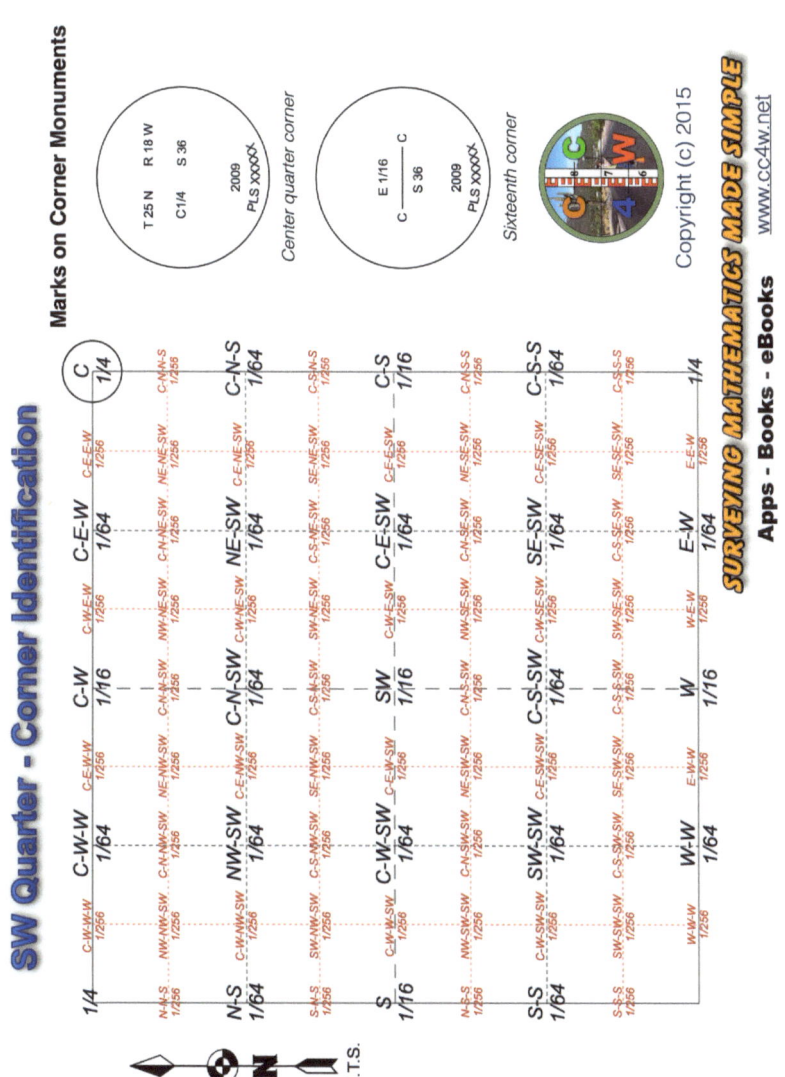

Charts & Graphs

THE LEGAL DESCRIPTION SYSTEM

The Legal Description System

There are Seven (7) types of legal descriptions

1. Bounds
2. Metes and Bounds
3. Public Land Survey System (PLSS)
4. Call for another document
5. "LY" descriptions
6. Strips
7. Lot and Block

(See Examples)

Copyright (c) 2015
www.cc4w.net

SURVEYING MATHEMATICS MADE SIMPLE
Apps - Books - eBooks

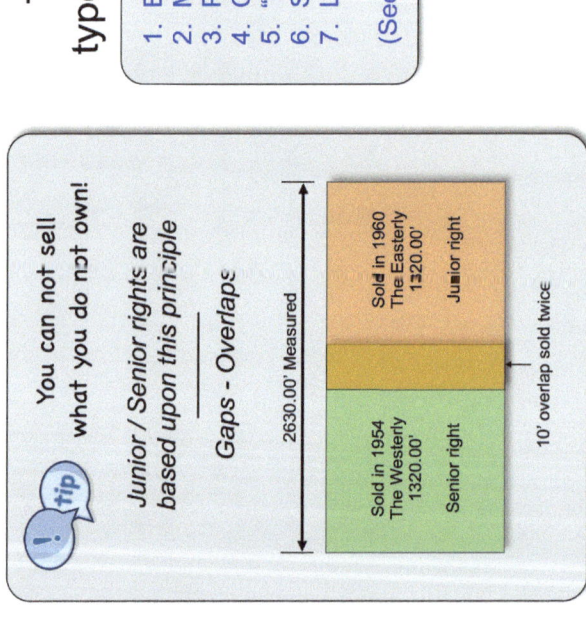

tip

You can not sell what you do not own!

Junior / Senior rights are based upon this principle

Gaps - Overlaps

2630.00' Measured

Sold in 1954
The Westerly
1320.00'
Senior right

Sold in 1960
The Easterly
1320.00'
Junior right

10' overlap sold twice

Charts & Graphs

PARTS OF A TYPICAL LEGAL DESCRIPTION

Parts of a typical legal description

Narrative forms:

Metes & Bounds -
Commencing at…to the Point of Beginning…(*courses*)…to said Point of Beginning;
Containing xxx Square Feet, xxx Acres;

LY descriptions -
The Easterly xxx feet of ….

Strip descriptions -
…lying on each side…to the Point of Termination.

Ties
From which the xxx corner bears xxx distance xxx.

Caption
The caption limits the title and/or the detailed description of the body. Some descriptions are contained all within the caption.

Body
The body expands on the Caption giving a more detailed description such as a metes and bounds description but is limited by the Caption.

Exceptions, Reservations, Clarifications
This is were you identify any exceptions, subject to, reservations, clarifications that aid in defining the legal description.

NOTE:
"The preferred method of writing descriptions is using the best type or combination of types and parts that will give the clearest and shortest description possible."
(Wattles "Writing Legal Descriptions")

Not all legal descriptions are written by Professional Land Surveyors

Copyright (c) 2015 www.cc4w.net

SURVEYING MATHEMATICS MADE SIMPLE
Apps - Books - eBooks

Charts & Graphs

BOUNDS

All of Tract 37 located in Section 12, Township 2 North, Range 4 East of the Gila and Salt River Meridian, Maricopa County, Arizona lying east of the Pima Drainage Canal east right of way line.

Copyright (c) 2015
www.cc4w.net

SURVEYING MATHEMATICS MADE SIMPLE
Apps - Books - eBooks

Bounds

N

East of Canal

Tr 37

Sec 12

Pima Drainage Canal

METES AND BOUNDS

Metes and Bounds

Those portions of the Southwest quarter of Section 25, the Southeast quarter of Section 26 and the Northeast quarter of Section 35, Township 41 North, Range 30 East of the Gila and Salt River Meridian, Apache County, Arizona, described as follows:

Commencing at a found 3 1/2 inch Bureau of Land Management (BLM) Department of Interior 2006 brass cap marking the common corner for Sections 25, 26, 35 and 36 of said Township, from which the quarter corner for Sections 26 and 35 of said Township bears South 89°21'45" West, 2638.40 feet being marked with a 3 1/2 inch BLM 2006 brass cap;

Thence South 89°21'45" West, 305.47 feet along the south line of said Section 26 to a point on the northwesterly right of way line of route U.S. 160 (Tuba City - Four Corners) being the **POINT OF BEGINNING**;

Thence South 34°09'26" West, 47.03 feet along said northwesterly right of way line;

Thence North 55°52'03" West, 599.82 feet;

Thence North 34°09'22" East 600.24 feet;

Thence South 55°49'38" East 596.45 feet to a point that lies on said northwesterly right of way line being the point of curvature of a non-tangent circular curve to the right having a radius of 7539.44 feet;

Thence along said northwesterly right of way line, from a local tangent bearing of South 32°06'34" West along said curve a distance of 225.60 feet through a central angle of 1°42'52" to the point of tangency;

Thence South 34°09'26" West, 327.23 feet continuing along said northwesterly right of way line to the **POINT OF BEGINNING**.

Said parcel of land containing 359,655 square feet (8.257 acres), more or less.

17

PLSS

Sections, Lots, Tracts, Mineral Claims, and Aliquot parts.

All of the Northeast quarter of the Southwest quarter of Section 12, Township 2 North, Range 4 East of the Gila and Salt River Meridian, Maricopa County, Arizona.

Subject to an Ingress and Egress easement across the southerly 10.00 feet thereof.

Copyright (c) 2015

www.cc4w.net

SURVEYING MATHEMATICS MADE SIMPLE
Apps - Books - eBooks

PLSS

N

Sec 12

NE4SW4

10' Ingress and Egress Esm't

CALL FOR ANOTHER DOCUMENT

Call for another document or instrument

All of that certain property as recorded in Docket 2005-0036478 of the Maricopa County Records, Maricopa County, Arizona.

Reference document must be recorded in a public repository such as the County Recorder's Office, State Department of Transportation, State Land Department, Bureau of Land Management, Tribal Agencies, etc.

Copyright (c) 2015

SURVEYING MATHEMATICS MADE SIMPLE
Apps - Books - eBooks
www.cc4w.net

Charts & Graphs

"LY"

The easterly 25.00 feet of the southerly 30.00 feet of Lot 3 of the Spectra Subdivision as recorded in Book 123 of Maps, page 45 of the Maricopa County Records, City of Tempe, Maricopa County, Arizona.

Subject to an Ingress and Egress easement over the west 10.00 feet thereof.

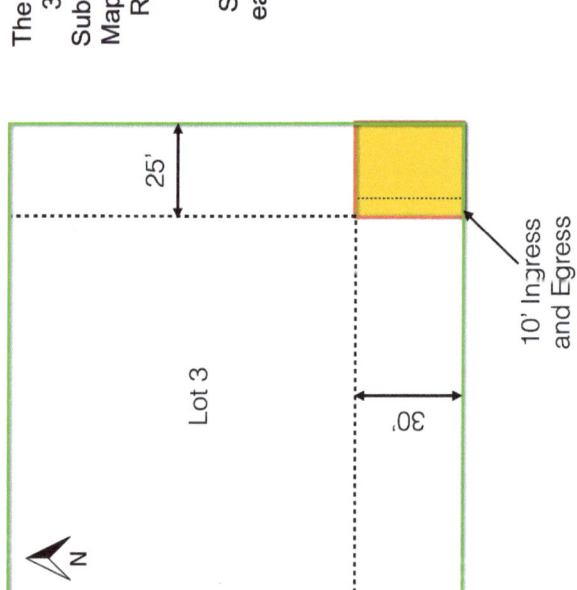

Charts & Graphs

STRIP

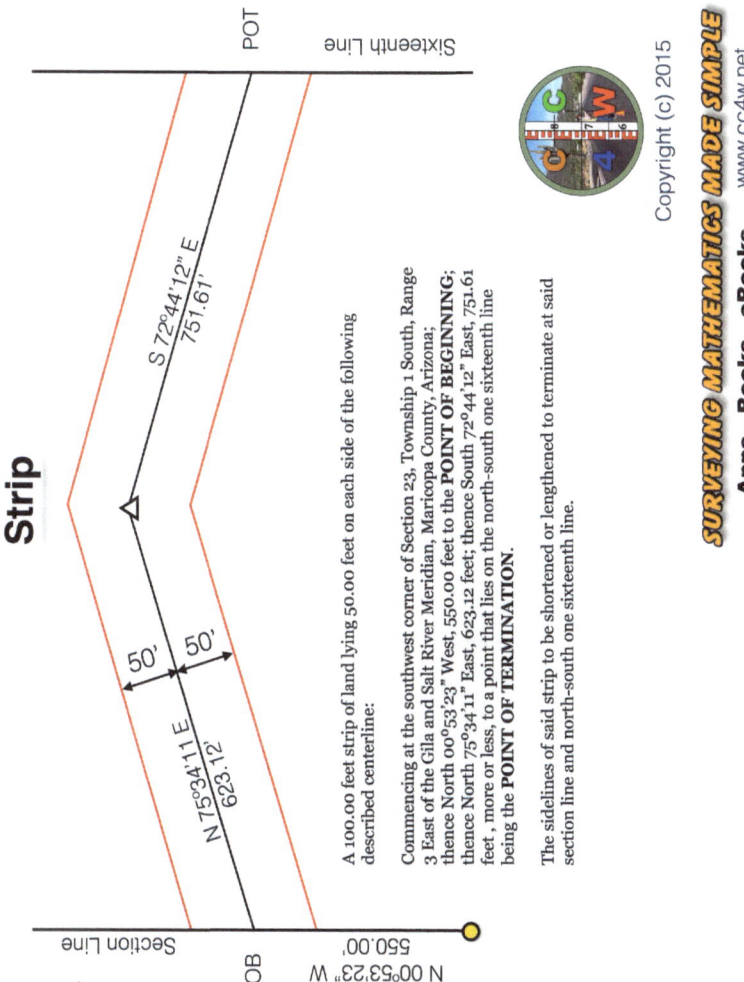

A 100.00 feet strip of land lying 50.00 feet on each side of the following described centerline:

Commencing at the southwest corner of Section 23, Township 1 South, Range 3 East of the Gila and Salt River Meridian, Maricopa County, Arizona; thence North 00°53'23" West, 550.00 feet to the **POINT OF BEGINNING**; thence North 75°34'11" East, 623.12 feet; thence South 72°44'12" East, 751.61 feet, more or less, to a point that lies on the north-south one sixteenth line being the **POINT OF TERMINATION**.

The sidelines of said strip to be shortened or lengthened to terminate at said section line and north-south one sixteenth line.

Charts & Graphs

LOT AND BLOCK

All of Lot 3 of Block 1 of the Spectra Subdivision as recorded in Book 123 of Maps, Page 45 of the Maricopa County Records, City of Tempe, Maricopa County, Arizona.

EXCEPT the north 10.00 feet thereof, ALSO EXCEPT the west 5.00 feet thereof.

Copyright (c) 2015

SURVEYING MATHEMATICS MADE SIMPLE
Apps - Books - eBooks
www.cc4w.net

Lot and Block

```
                    Lot 4
       ┌──────────┬──────────┐
       │          │  Lot 5   │
       │   10'    ├──────────┤
       │  Lot 3   │  Lot 6   │
       │          │          │
Block 1│  5'      │My Street │
       │  Lot 2   │  Lot 7   │
       │          │          │
   N   │  Lot 1   │  Lot 8   │
       └──────────┴──────────┘
```

22

Charts & Graphs

TANGENT DEFLECTION

This graph illustrates the point at which a tangent bearing should be shown to one decimal place for the seconds to reduce the rounding affect for long tangent lines. The threshold is 0.015' for the rounding affect.

Tangent line one:
N34°23'45.4"E 12,500.00'
Go up the Y axis to 12,500.
Go across the X axis to .4".
The point falls on the right side of the line graph. The .4" should be shown.

Tangent line two:
S12°23'22.6W 5000.00'
Up to 5000 then right to .4".
In this case work backwards from 23.0". The nearest second should be shown such as S12°23'23"W.

Tangent Deflection

Tangent Deflection

$f(x) = 0.015 / \tan(x)$

Show tenths of a second for tangent bearing

Show to nearest second for tangent bearing

x = Tenths of a second

f(x) = Distance in thousands of feet

Copyright (c) 2015

SURVEYING MATHEMATICS MADE SIMPLE
Apps - Books - eBooks
www.cc4w.net

Charts & Graphs

DEGREE OF CURVATURE

This graph illustrates the function between a 100' chord distance and it's relationship to the arc length.

Example:

X = 6 (Degree of Curvature)
Y = 0.046 feet

The arc length is 0.046 feet longer than the 100' chord.

Copyright (c) 2015

www.cc4w.net

SURVEYING MATHEMATICS MADE SIMPLE
Apps - Books - eBooks

Degree of Curvature

Chord vs Arc function

$f(x) = x * 50 / \sin(x/2) * \pi / 180 - 100$

x = Degree of Curvature (Chord Def)

f(x) = Arc distance - Chord distance

ABOUT THE AUTHOR

Jim Crume P.L.S., M.S., CFedS

My land surveying career began several decades ago while attending Albuquerque Technical Vocational Institute in New Mexico and has traversed many states such as Alaska, Arizona, Utah and Wyoming. I am a Professional Land Surveyor in Arizona, Utah and Wyoming. I am an appointed United States Mineral Surveyor and a Bureau of Land Management (BLM) Certified Federal Surveyor. I have many years of computer programming experience related to surveying.

This ebook is dedicated to the many individuals that have helped shape my career. Especially my wife Cindy. She has been my biggest supporter. She has been my instrument person, accountant, advisor and my best friend. Without her, I would not be the professional I am today. Cindy, thank you very much.

Other titles by this author:

http://www.cc4w.net/ebooks.html

Follow us on Facebook

Books available on amazon.com

SURVEYING MATHEMATICS MADE SIMPLE
MATH-SERIES TRAINING AND REFERENCE BOOKS / APPS

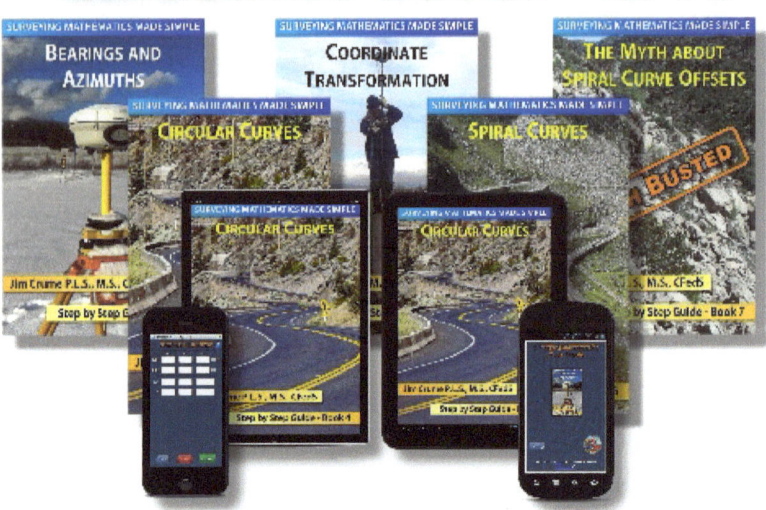

Printed - Digital - Apps
Many Titles to choose from.
www.cc4w.net

A **New** Math-Series of books with useful formulas, helpful hints and easy to follow step by step instructions.

www.facebook.com/surveyingmathematics

Digital and **Printed Editions** Math-Series Training and Reference Books. Designed and written by Surveyors for Surveyors, Land Surveyors in Training, Engineers, Engineers in Training and aspiring Students.

www.ingramcontent.com/pod-product-compliance
Lightning Source LLC
Chambersburg PA
CBHW041616180526
45159CB00002BC/886